What's Your Poo Telling You?

A poopourri of activities featuring POO, PEE, and FARTS!

ACTIVITY BOOK

CHRONICLE BOOKS
SAN FRANCISCO

978-0-8118-7457-1

Manufactured in China

Design by Sarah Pulver

10 9 8 7 6 5 4 3

Chronicle Books LLC
680 Second Street
San Francisco, CA 94107
www.chroniclebooks.com

Where All the Magic Happens

Identify the organs that make up your digestive and urinary tract systems. Without them, you wouldn't be able to poo, pee, or fart!

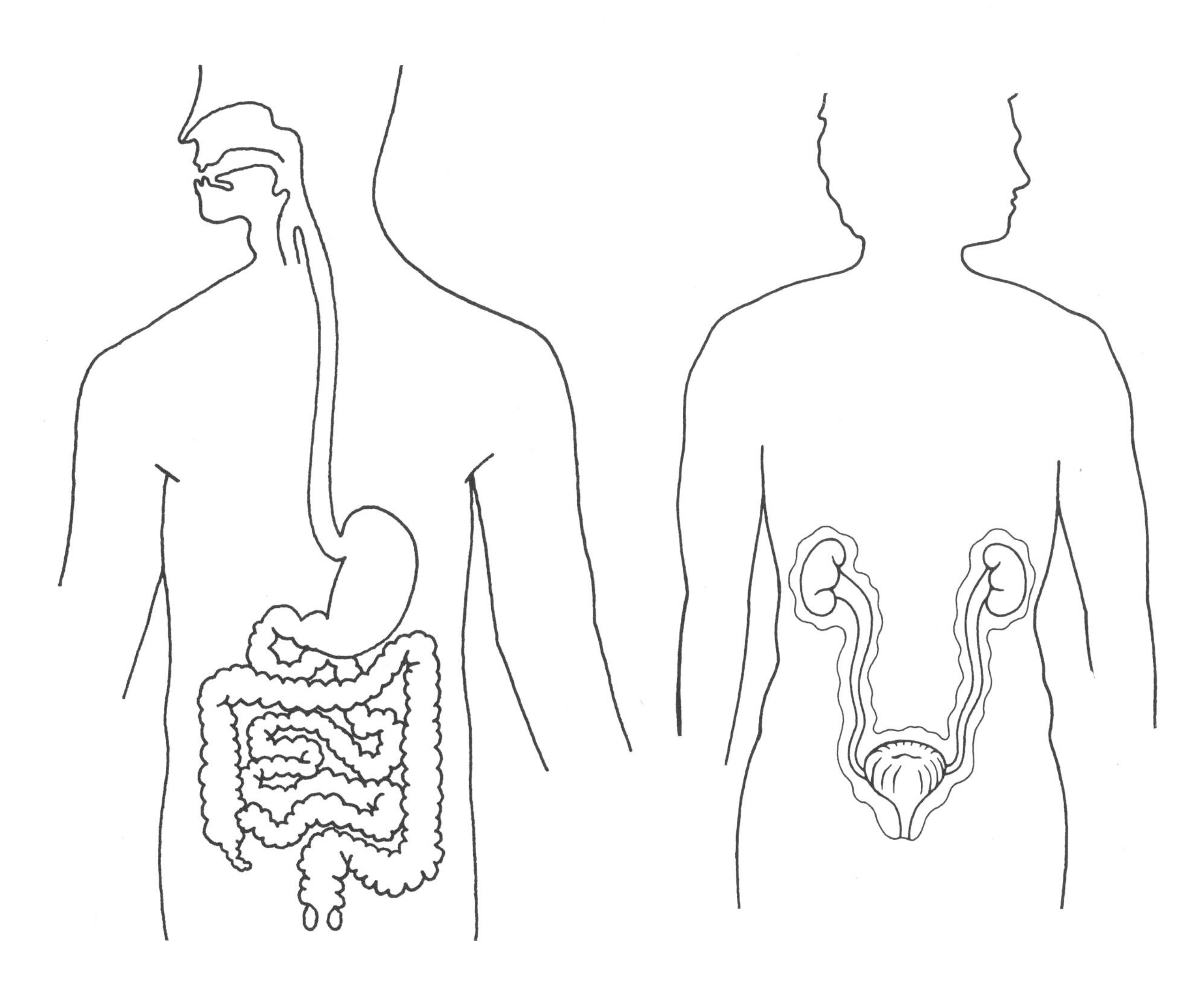

STOMACH
SMALL INTESTINE
LARGE INTESTINE (COLON)
RECTUM

RIGHT KIDNEY
LEFT KIDNEY
URINARY BLADDER
URETER
URETHRA

See answers on page 31.

A Poopourri

Color in these poos according to their descriptions.

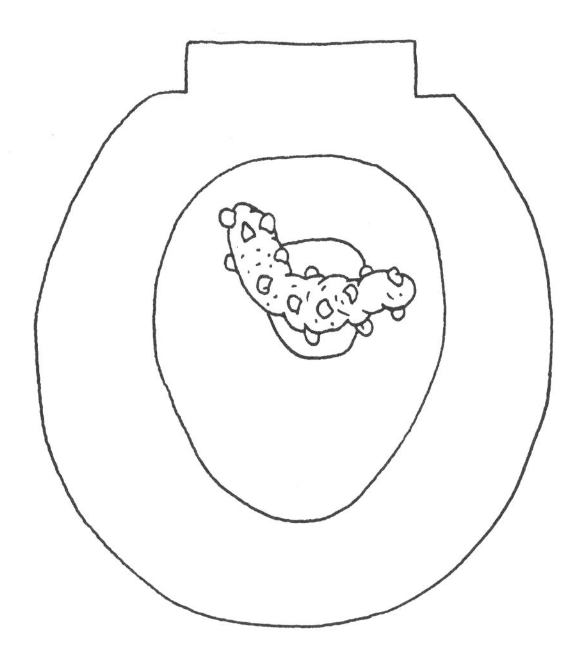

Deja Poo: includes familiar portions of a recent meal embedded in it, often containing pieces of vegetables (most notoriously corn); can include a potpourri of colors.

Camouflage Poo: multi-toned poo (varying shades of black, brown, and green); looks like a mosaic of diverse excrement from different sources and from assorted meals—as if many pieces of poo were forced together.

Soft Serve: more dense than diarrhea, but softer than normal poo, this solid yet amorphous turd comes out in one smooth steady flowing motion.

The Snake: long, thin, and windy defecation; can contort itself into a variety of different shapes and sizes.

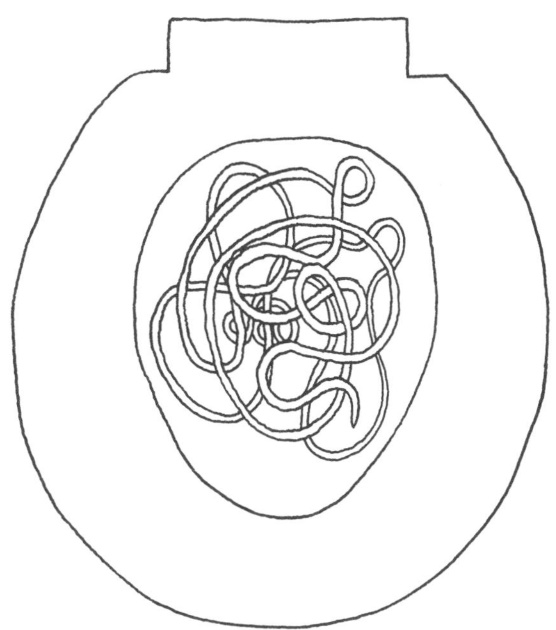

Pebble Poo: hard, disjointed poos, like a handful of pebbles.

The Chinese Star: hard, angular bowel movement; the defining characteristic is the excruciatingly painful sensation as if the rectum is being torn apart from the inside as this turd exists the body.

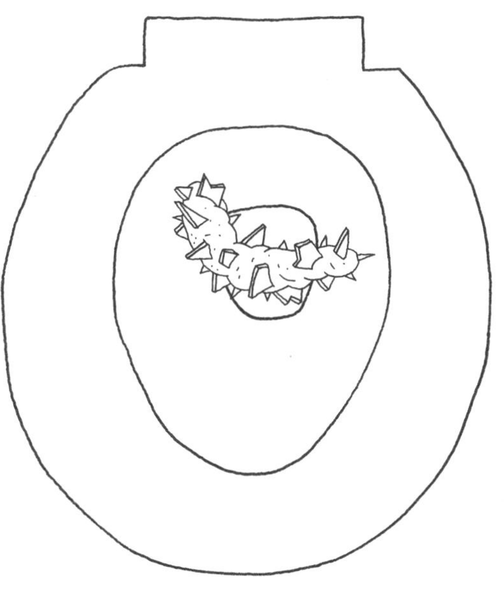

To sit, or to squat, that is the question.

Number Three

a.k.a. Butt Piss, Liquid Poo, Montezuma's Revenge, Poo Stew, Turd Tea

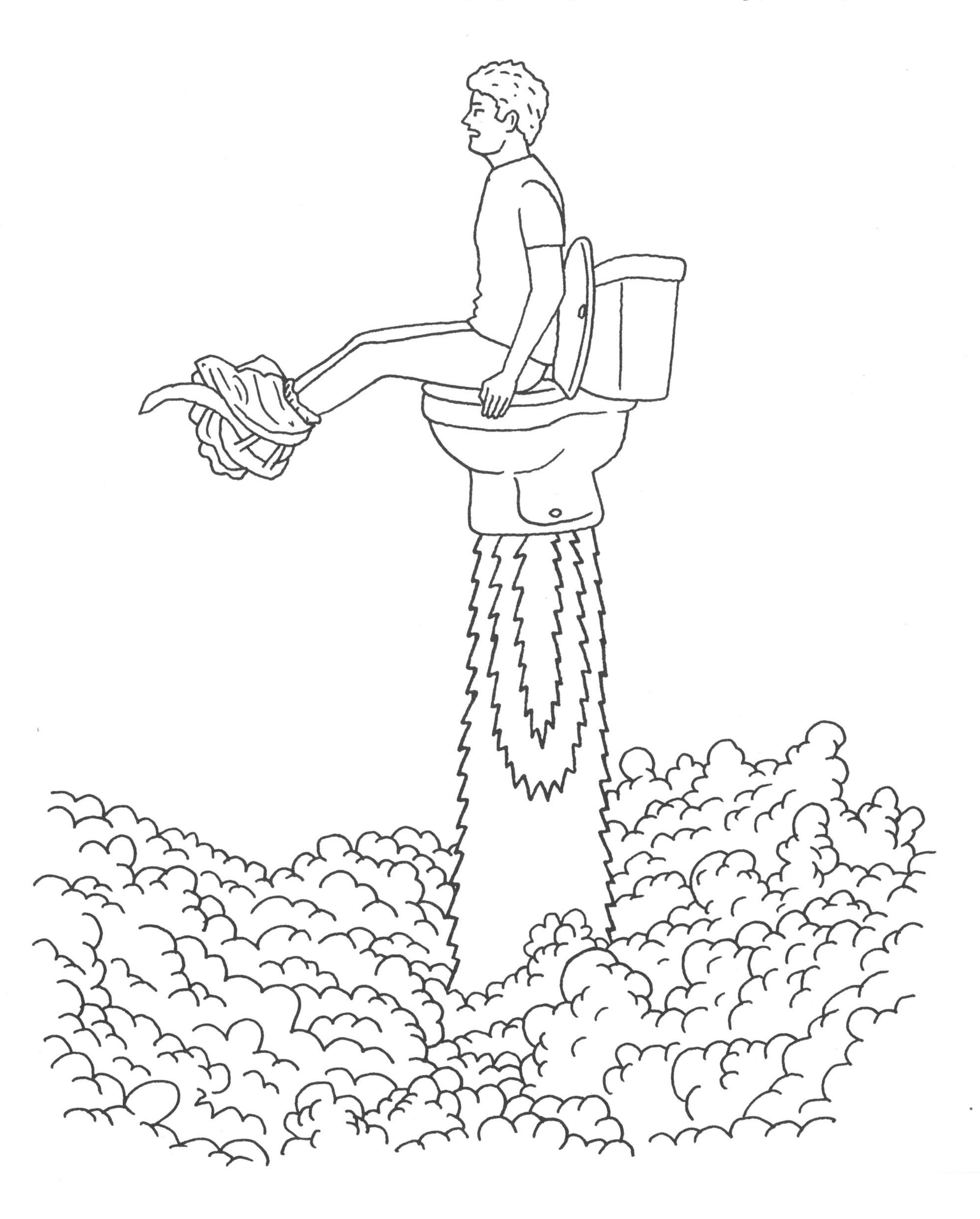

D.A.D.S., the Day-After-Drinking Stool

Had a little too much to drink last night? Create your own post-party bathroom scene.

The Honeymoon's-Over Poo

Pooing freely in your partner's presence is a sure sign that the honeymoon's over.

Take cover . . . it's a poo-nami!

Word Find

Words can be found up, down, forwards, backwards, and diagonally.

BLADDER	DRIP	PEBBLE	SHART
BUCKSHOT	ENEMA	PEEPHORIA	SKIDMARK
COLONOSCOPY	FLATULENCE	POO	SNAKE
CRAPPER	FLOATER	POONAMI	SQUAT
DIARRHEA	LAXATIVE	RECTUM	STOOL
DIGESTION	NUGGET	SEEPAGE	STREAK

E	E	P	R	P	S	O	R	P	P	V	G	J	V	Y
F	L	A	T	U	L	E	N	C	E	K	I	I	P	S
X	N	B	X	V	D	Y	B	D	E	F	X	O	Q	K
C	U	G	B	D	Y	U	M	C	P	A	C	U	R	R
K	I	R	A	E	C	F	K	R	H	S	A	C	Z	A
S	F	L	P	K	P	A	L	O	O	T	S	F	X	M
C	B	W	S	C	E	F	O	N	R	J	D	T	C	D
U	A	H	F	R	Z	B	O	E	I	L	I	L	E	I
E	O	F	T	L	B	L	C	R	A	P	P	E	R	K
T	V	S	Q	T	O	A	E	H	R	R	A	I	D	S
D	P	I	H	C	P	A	T	V	P	P	M	Z	V	N
Q	I	I	T	O	S	L	T	Q	R	O	U	E	O	A
X	R	G	R	A	I	H	S	E	E	O	T	C	B	K
G	D	D	E	D	X	E	A	K	R	N	C	J	W	E
B	K	E	I	S	E	A	L	R	R	A	E	Q	D	K
O	R	F	S	P	T	C	L	W	T	M	R	J	H	O
M	N	F	A	A	O	I	Y	L	U	I	Q	X	F	O
Q	Q	G	M	A	E	K	O	T	E	G	G	U	N	P
Y	E	N	E	M	A	E	B	N	H	F	G	S	W	F
J	D	D	R	C	V	X	G	D	T	E	P	Z	X	C

See answers on page 31.

Toilet Time

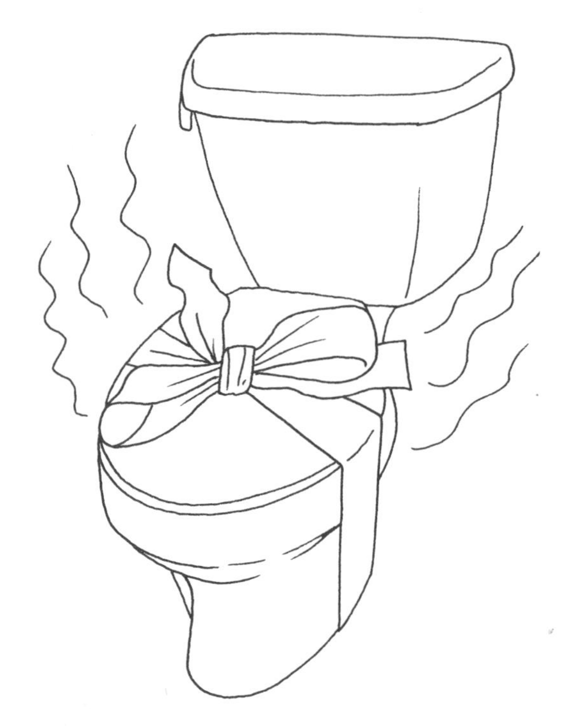

Gift Poo: turds that people leave behind in toilets without flushing.

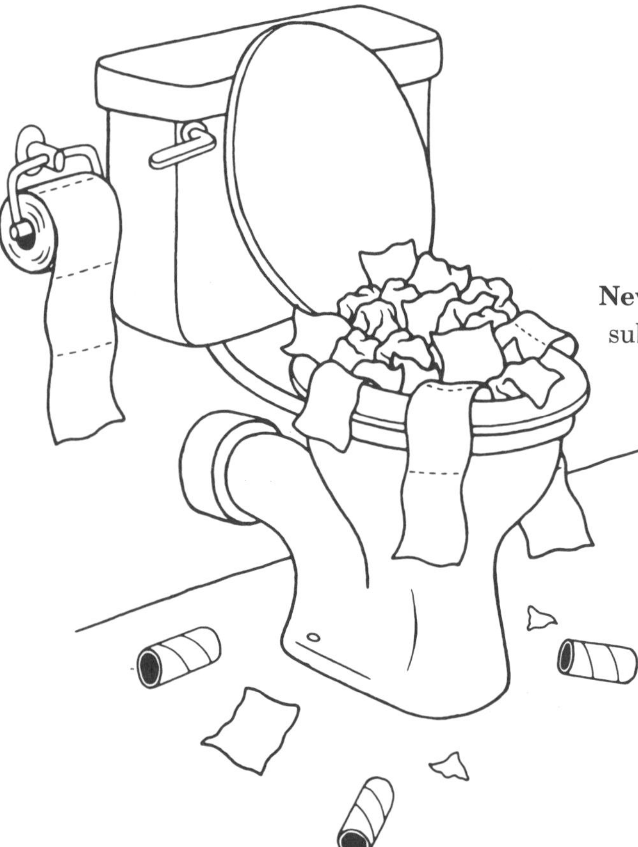

Never-ending wipe poo: stickier-than-usual substance in greater-than-expected quantity.

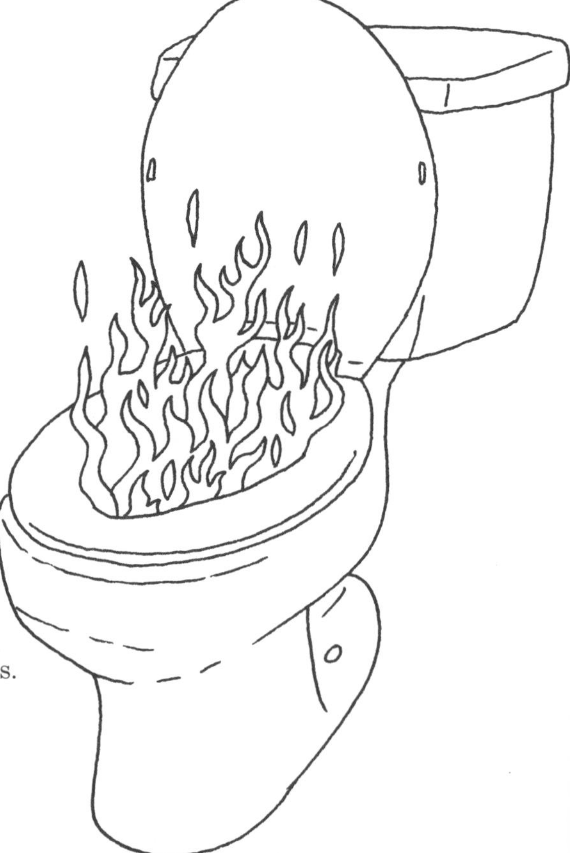

Ring of Fire: burning sensation that rips through anus.

Connect the dots and find out what surprise awaits you!

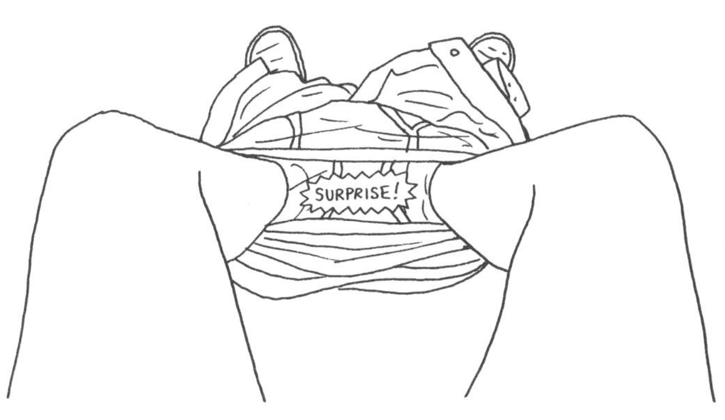

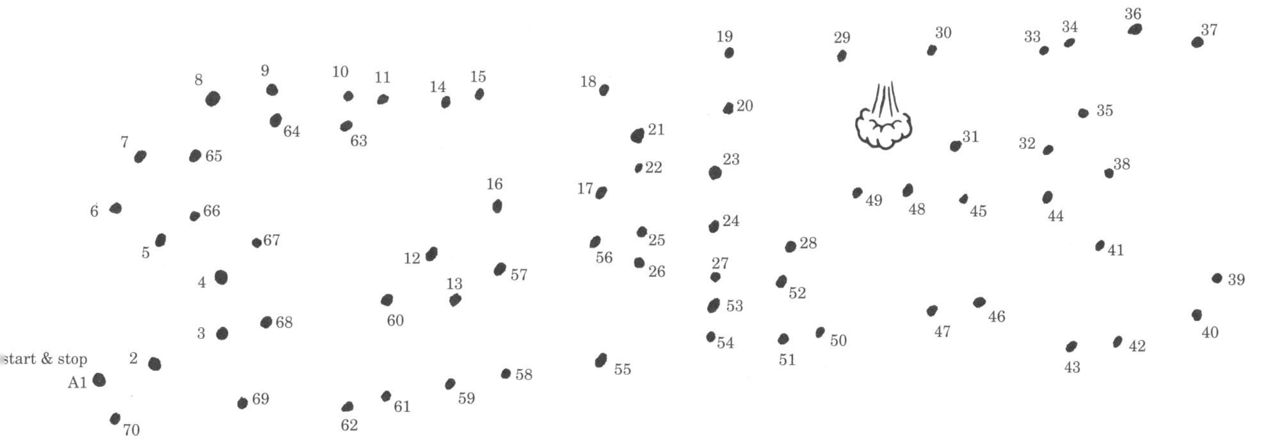

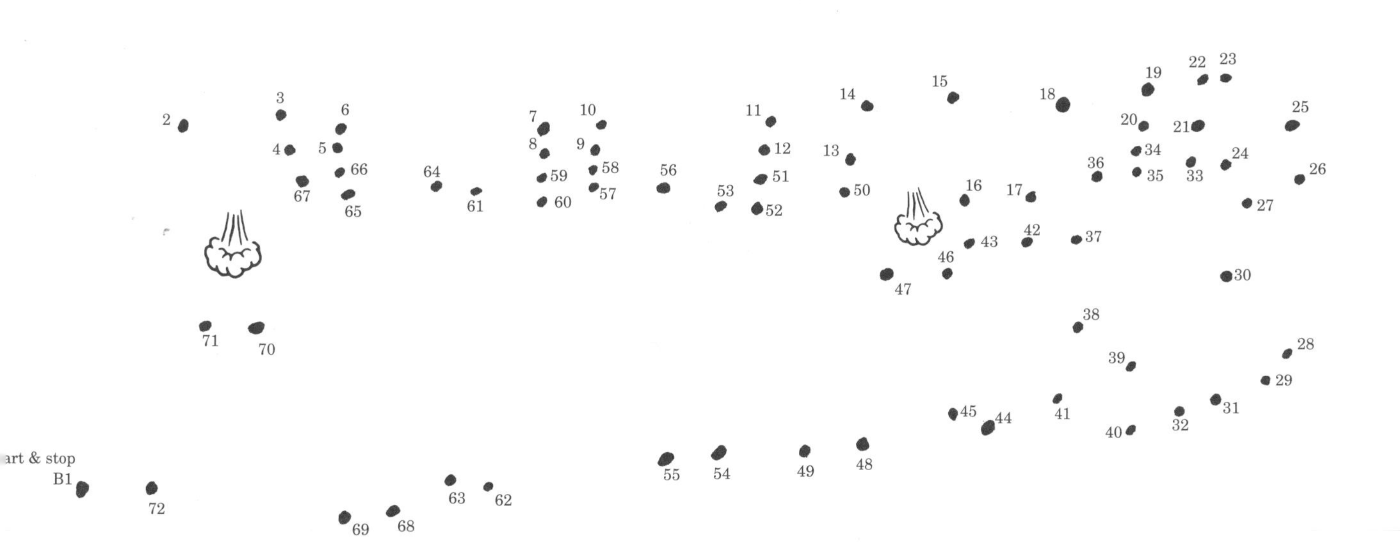

Multiple Choice

1. The word "poop" comes from the word poupen. What does poupen mean?
   - a. feces
   - b. pee
   - c. fart

2. Farts consist of:
   - a. nitrogen
   - b. carbon dioxide
   - c. hydrogen
   - d. methane
   - e. all of the above

3. Poo is mostly:
   - a. digested food
   - b. water
   - c. bacteria

4. The town 'Badfart' is found in which country?
   - a. United States
   - b. Sweden
   - c. Hungary

5. What is the longest documented human poo?
   - a. 5 feet
   - b. 10 feet
   - c. 26 feet

6. Which of the following is a cause of albino poo?
   - a. drinking lots of milk
   - b. eating lots of chalk
   - c. bile duct blockage

See answers on page 31.

Poodoku

Place the nine given letters once in each column, row, and three-by-three box. When done, read the first row for a special poo-related word prize.

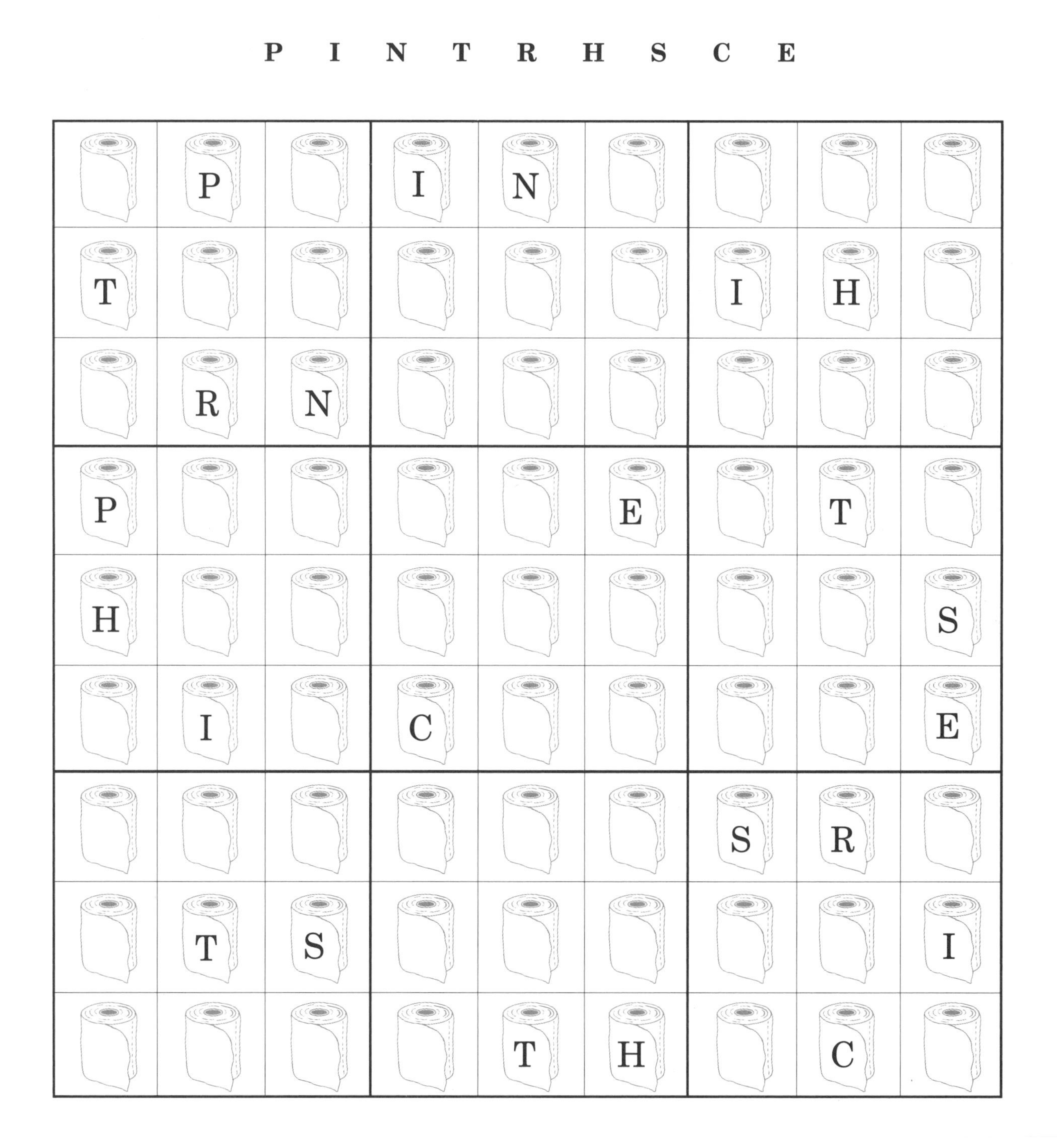

See answers on page 31.

Vitamin P

Put on your sunglasses and pull out your fluorescent crayons!

Dr Stool says: Bright yellow or orange urine is virtually always caused by the ingestion of vitamins and medications.

Why can't men pee straight?

Pee-phoria

Sometimes we are forced to hold in our pee, and what begins as a small urge rapidly becomes an unbearable discomfort. Peephoria is the result of finally letting it go!

The Hot Box

Word Scramble

Unscramble the words to find out the answers to these poo-rific facts:

1. Toilet paper was invented in this country. **I H A N C**

2. This animal eats its own poo. **B A T B R I**

3. This animal doesn't poo at all when hibernating. **A R B E**

4. The earliest documented use of enemas was in 1500 B.C. in this country. **Y T P E G**

5. Seventeenth-century alchemists believed that the yellow color of pee was due to the presence of this element in urine. **D O L G**

see answers on page 31.

Monster Poo

Add cracks to the walls to show just how monstrous this poo really is!

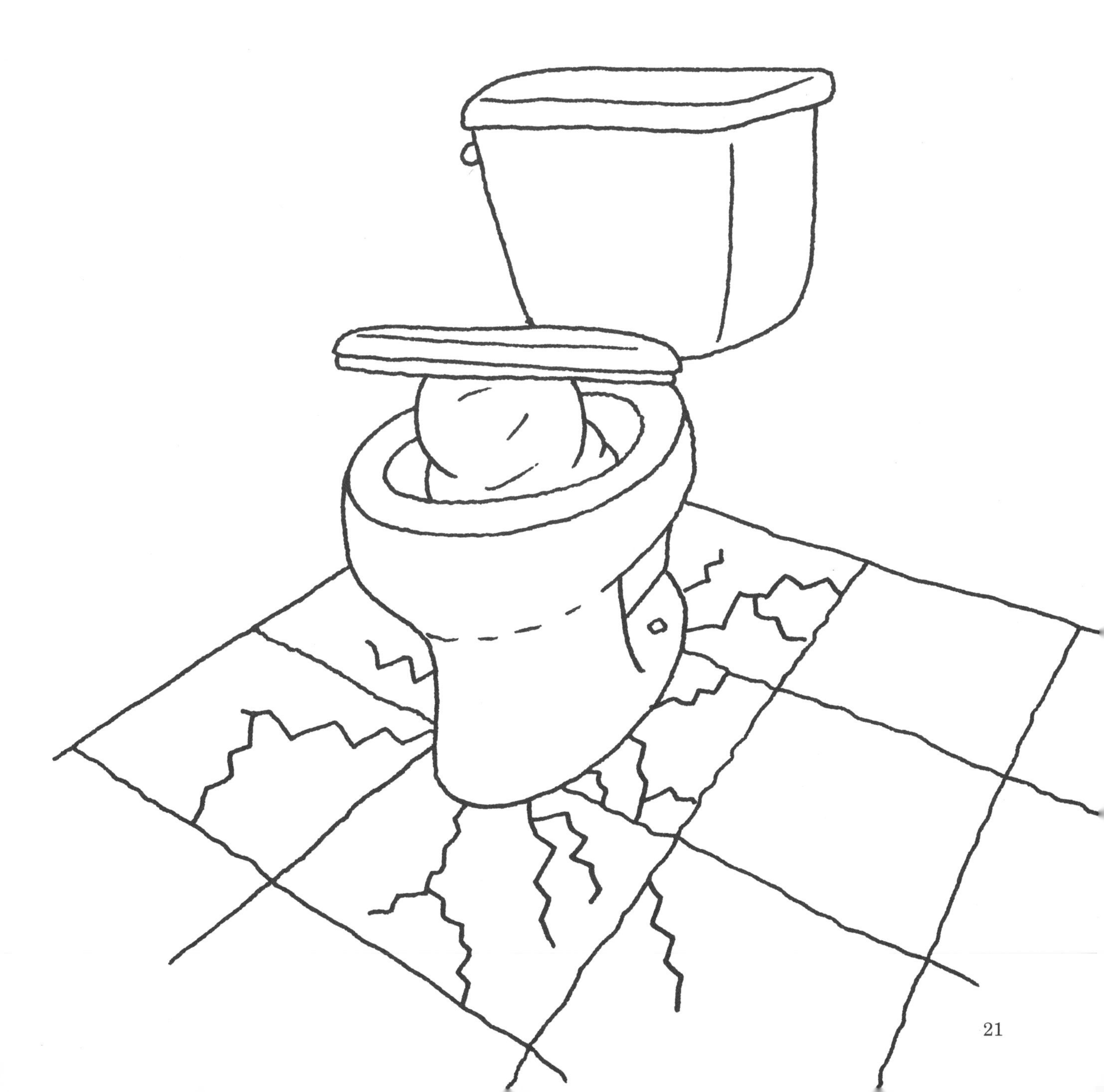

The Explosion

Characterized by its deafening sound, the “explosion” is likened by some to the roar of a lion.

Morning Thunder

Dr. Stool says: Morning Thunder is the result of an increase in the colon's muscular contractions which occur upon awakening.

Silent-but-Deadly Spot the Difference

Find all 10 differences between these two Silent-but-Deadly scenarios.

See answers on page 31.

True or False

Peeing on a jellyfish sting will relieve the pain?

See answer on page 31.

True or False

Sitting on the toilet for too long causes hemorrhoids?

See answer on page 31.

True or False

It is possible to light farts on fire?

See answer on page 31.

A-Mazing Poo

See answer on page 31.

Draw your own poo!

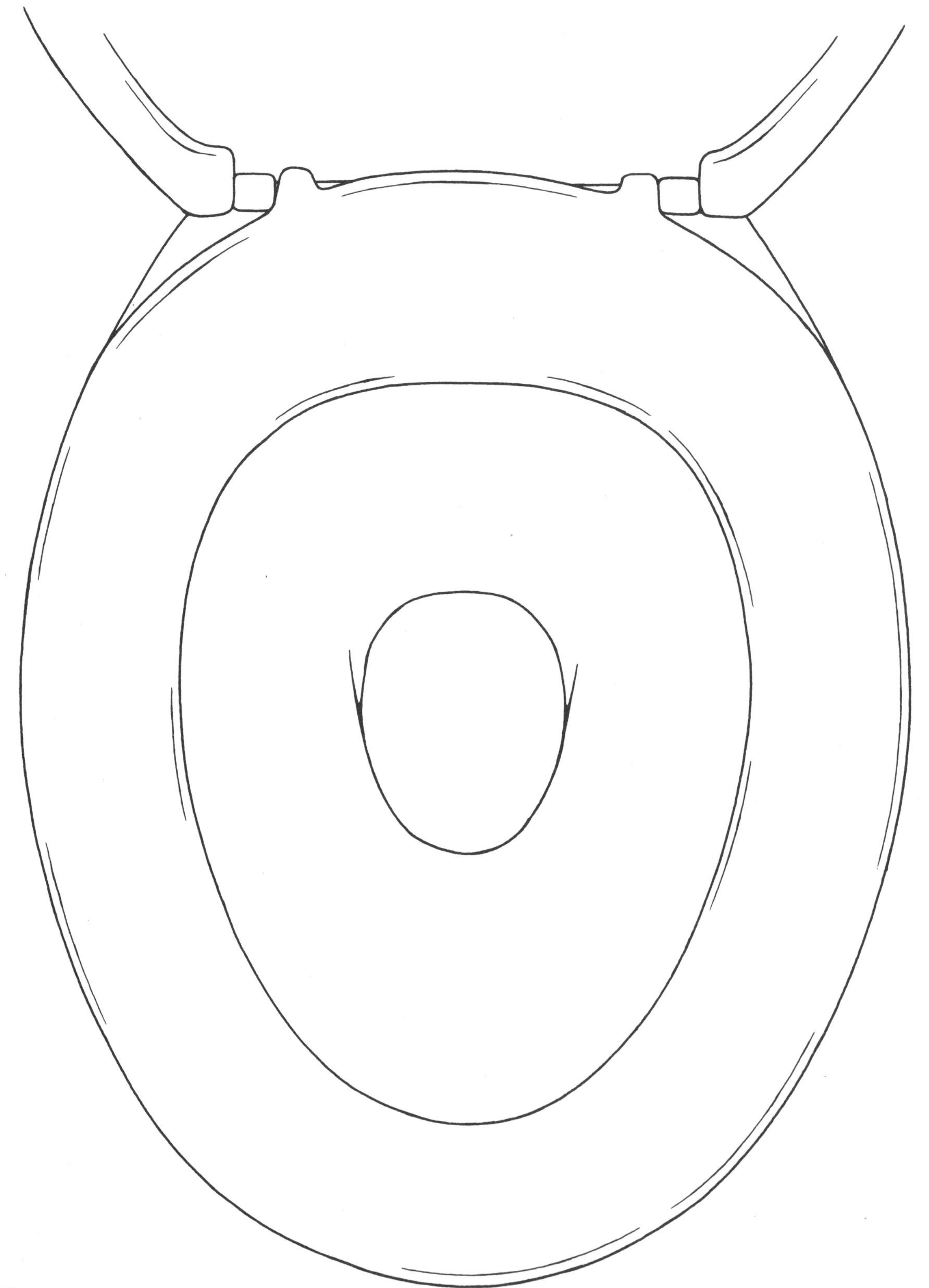

Where All the Magic Happens:

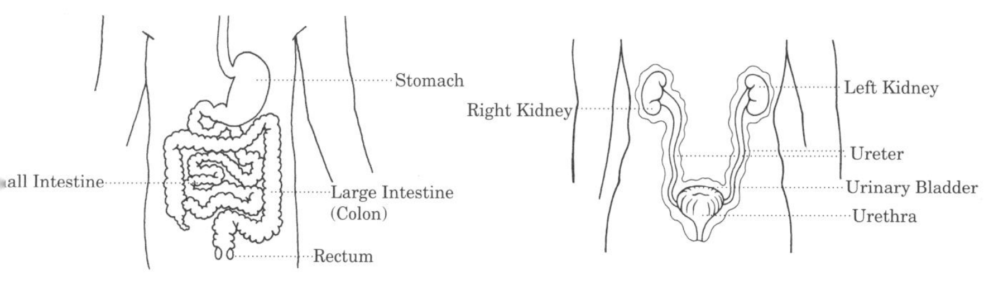

Word Find:

E	E	P	R	P	S	O	R	P	P	V	G	J	V	Y
F	L	A	T	U	L	E	N	C	E	K	I	I	P	S
X	N	B	X	V	D	Y	B	D	E	F	X	O	Q	K
C	U	G	B	D	Y	U	M	C	P	A	C	U	R	R
K	I	R	A	E	C	F	K	R	H	S	A	C	Z	A
S	F	L	P	K	P	A	L	O	O	T	S	F	X	M
C	B	W	S	C	E	F	O	N	R	J	D	T	C	D
U	A	H	F	R	Z	B	O	E	I	L	I	L	E	I
E	O	F	T	L	B	L	C	R	A	P	P	E	R	K
T	V	S	Q	T	O	A	E	H	R	R	A	I	D	S
D	P	I	H	C	P	A	T	V	P	P	M	Z	V	N
Q	I	I	T	O	S	L	T	Q	R	O	U	E	O	A
X	R	G	R	A	I	H	S	E	E	O	T	C	B	K
G	D	D	E	D	X	E	A	K	R	N	C	J	W	E
B	K	E	I	S	E	A	L	R	R	A	E	Q	D	K
O	R	F	S	P	T	C	L	W	T	M	R	J	H	O
M	N	F	A	A	O	I	Y	L	U	I	Q	X	F	O
Q	Q	G	M	A	E	K	O	T	E	G	G	U	N	P
Y	E	N	E	M	A	E	B	N	H	F	G	S	W	F
J	D	D	R	C	V	X	G	D	T	E	P	Z	X	C

Multiple Choice:

1. c: fart
2. e: all of the above
3. b: water
4. b: Sweden
5. c: 26 feet
6. c: bile duct blockage

Poodoku:

S	P	H	I	N	C	T	E	R
T	C	E	P	S	R	I	H	N
I	R	N	H	E	T	P	S	C
P	S	C	R	I	E	N	T	H
H	E	R	T	P	N	C	I	S
N	I	T	C	H	S	R	P	E
E	H	P	N	C	I	S	R	T
C	T	S	E	R	P	H	N	I
R	N	I	S	T	H	E	C	P

A-mazing Poo:

Word Scramble:

1. China
2. rabbit
3. bear
4. Egypt
5. gold

True or False:

Page 26: False
Page 27: True
Page 28: True

Silent-but-Deadly Spot the Difference:

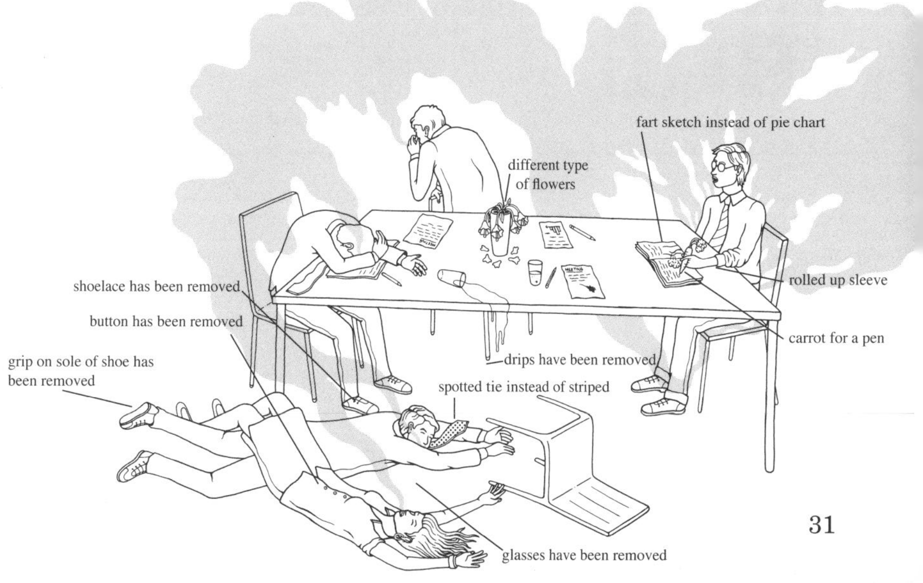

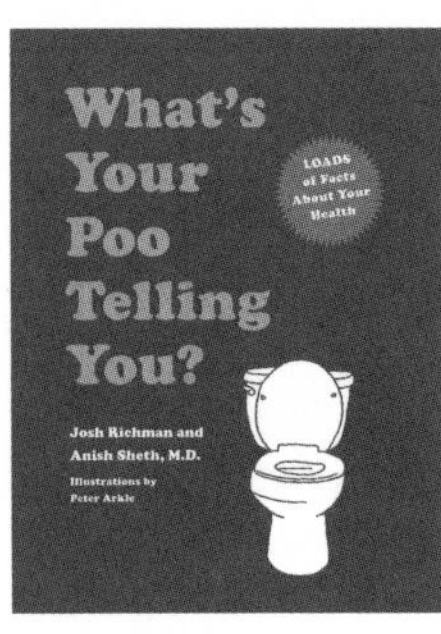

What's Your Poo Telling You?

With universal appeal—after all, everyone poops—this straightforward, illustrated description of two dozen dookies (each with a medical explanation written by a doctor) details what one can learn about health and well-being by studying what's in the bowl. A floater? It's probably due to a build-up of gas. Now think back on last night's dinner—a burrito perhaps? All the greatest hits are here: The Log Jam, The Glass Shard, The Hanging Chad . . . the list goes on. Sidebars with interesting factoids, 60 euphemisms for number 2, and unusual case histories all make this the ultimate bathroom reader.

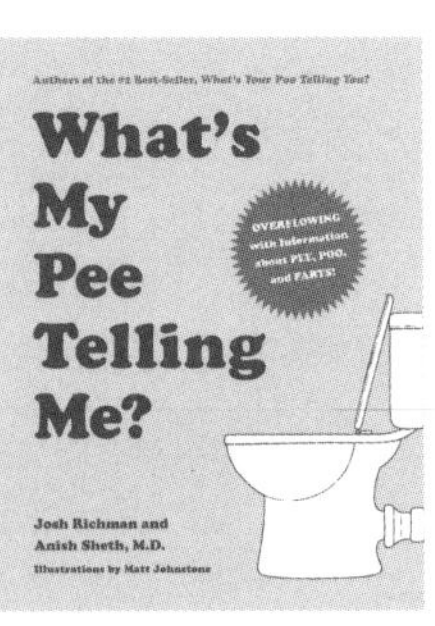

What's My Pee Telling Me?

What goes in must come out. It's that simple. But what does it all mean? Therein lies the mystery—and the key to your health and happiness. In this entertaining and fact-filled guide, the authors of the best-selling *What's Your Poo Telling You?* expand their probing inquiry into the workings of the human body to reveal the secrets and splendors of farts and pee, as well as more about their inevitable companion, poo. In the shocking and informative final section, the authors explode a variety of popular myths about the gastrointestinal tract.

Poo Log

Finally, what every bathroom has been waiting for—the *Poo Log*, a journal for recording and studying the wondrous uniqueness of each bowel movement. With an extensive glossary, handy reference checklists, and interesting nuggets throughout, this journal makes every trip to the can an e-loo-cidating experience. Who knew one could learn so much from poo?